THE SOMEWHAT UNORTHODOX GUIDE TO GETTING 'A' IN MATH EXAMS.

A guide for students, teachers, and parents.

Chew Sze Chong

THE SOMEWHAT UNORTHODOX GUIDE TO GETTING 'A' IN MATH EXAMS.

A guide for students, teachers, and parents.

www.h2math.com

Chew Sze Chong

INTRODUCTION

Since 2013 I have been tutoring students one to one for Cambridge 'A' level as well as International Baccalaureate SL and HL Mathematics. I learned over the years that the level of success of students in mathematics must do mainly with how they approach their study and their exam preparation strategies. I would like to share them with you in this book.

This book is mainly for students with the desire to get a Grade A in mathematics.

I don't want to waste your time because you would be better off spending time doing math questions, so I'm getting straight to the point in most cases and I'm not going to make lengthy arguments for what I realized through actual experience, since these are not mainstream theories of learning, therefore I call it the unorthodox guide, nevertheless it worked for me and my top students.

If you like to chat with me, you can write to me at genie10001@gmail.com or WhatsApp to +65 81808611. I reside in Singapore and I'm tutoring students for Advance Level mathematics for both the International Baccalaureate and the Cambridge Syllabus.

This is my website: www.h2math.com

MINDSET

Read through this chapter to get the mindset that will help you to do well in Mathematics.

The most successful students I have taught look at getting A in mathematics and becoming good at mathematics as an end and not the means to something else. It is NOT a means to enter the university, or get a good career, or impress someone. They want to get A in mathematics because they want to get A in mathematics and want to be good at mathematics. If you think like that, it is easier for you to get an A. You must enjoy the idea of getting A in math.

It is not difficult to understand why these students are exceptionally successful. Do you ever ask yourself what is the purpose of playing an addictive game or reading an addictive book or watching a drama? Are these things good for you or serve some higher purpose? Not necessary, but you enjoy them, that is why you can persist and do them well. You can excel in an activity because you are interested in the activity itself or if you know it is an important part of helping you achieve an important goal of yours.

However, there is indeed a good purpose for training in mathematics. When mathematics is learned the right way, it trains your cognitive ability like abstraction, memory, logical reasonings, structured thinking as well as problem–solving skills, attitude, process, and habits. These skills also become the foundation necessary for learning and mastering other practical disciplines such as engineering, computing, economics, business, and etcetera. I was able to learn coding on my own without a tutor and getting A for my business finance and economic modules in my MBA course with NUS because of my good foundation in Mathematics.

PRACTICE UNTIL YOU GET IT, NOT TO FULFILL HOURS

Students ask me how many hours they should study in a day? When I was a student, I gave priority to the learning of mathematics. I make sure I finish every single math question. If there is a formula, I study until I retain it in memory, if it takes days or week, I spend that amount of time. If there a difficult question, I work on it until I crack it. I may think on hours or days or months, flip and flip textbooks, reread the question, check my errors for as many as a few hundred times and would not give up until I get how it works.

And I will make sure I redo a difficult question again after a few days to make sure I'm not depending on short term memory to do the question. For my honors years thesis, I think of the problem when I'm eating breakfast, eating lunch, lying on the bed before sleep for almost a month before I finally understood and able to explain the result in the thesis.

My student once commented that I'm like an Encyclopedia for H2 mathematics because when I'm answering their question, I do not refer to formula sheets, textbooks or answers on the syllabus. Everything is right inside my mind.

You need to have the desire to practice until you get it and retain it and able to retrieve it from memory effortlessly.

But I must tell you, this approach is not for getting straight A in all subjects, I got A for all my Math subjects from Primary School to JC but not for all other subjects. To get straight As you may need to balance the time spent on different subjects and you will need to engage tutors for your weaker subjects. Good tutors cut short the time you need to understand concepts and impart neat and quick ways of doing things that they picked up over years. I never had a private tutor for any subject, I picked up and invented shortcuts through the decades of learning and teaching.

Therefore, the short answer to how many hours you should study in a day is: as many hours as you need to solve all the math questions you have and remember all the things you need to remember. You must be prepared to give your all. You must leave no stones unturned before you stop. This means doing correction for any mistakes you have in tests and exams and making sure you can do every single question in the assignment. You must go back regularly to turn the stones again. This means revising chapters you have not touched for a few weeks to make sure you have not forgotten them.

Do take breaks when you are exhausted. Personally, I could not let go of a challenging math problem I encountered is solved. I will take breaks from the pen and paper to walk around, but my mind is still turning around the problem. As soon as I get some inspiration, I will be back on the pen and paper to try it out. My taking break is different from how some of my students take a break, they take a break from the question and don't go back to it until days later. When you are passionate in something and do more than your peers, your result is likely to be exceptional.

If you need time for other subjects but can't find enough then you may need a private tutor who can answer your questions and guide you to understand things faster. It is better to get a tutor that makes himself always available. I meet my students once or twice a week depending on demand. My students can WhatsApp me their questions anytime and I will revert to them within a day with some supply of hints or concepts which they could apply to solve the question.

If you want to engage a tutor online and we happen to be in the same time zone, you can chat me up using WhatsApp to +65 81808611, I live in Singapore and we are famous for having one of the most difficult version of Cambridge 'O' and 'A' level math exam papers in the world. Singaporean students usually came in the top in International Math Competition.

DOING CORRECTIONS IMMEDIATELY

After teaching more than 50 students and referring to my own learning experience, I can predict a student's level of success based on how they regard corrections. It is important to do the correction immediately when you receive an exam, test paper or homework question that is marked wrong. When I received homework, test or exam paper when I was a student, I will quickly scan the questions to see the ones I got wrong. Then I will immediately launch into doing correction by myself, without referring to the solution, because I want to make sure I can do it right, by myself.

A successful student will immediately do correction when I tell them something is wrong. They will redo the question neatly by the side to make sure they can get it by themselves.

I also have students who are just contented to leave it like it is and want to move on to the next question without bothering to do the correction. These second type of students are usually less successful.

The difference in attitude towards doing correction reflect essential differences in attitude and habits to learning which limit their level of success. The students who do correction wants to ensure that they understood correctly and wants to ensure that they have written down something for doing revision later. They want to leave no stones unturned. This attitude ensures they are successful.

The students who do not do correction have one or more of these believe that lead them to become unsuccessful:

1)   They just want to be exposed to as many questions as they can but do not bother to make sure they can do every question they encounter. Therefore, when they encounter a similar question later, they think of skipping it or do not have the confidence to solve it because they have no prior experience of solving it by themselves.

2)   They want to improve their confidence by doing more easy questions. They don't want to dwell on their mistakes. They hope that those questions will not

appear in exams. This kind of thinking will ensure the students are unsuccessful in the exam because firstly, teachers like to set difficult questions in exams to test their student's level of understanding, not easy questions. Secondly, if the question is easy, every student can do it, meaning your score will always be average or below average if you focus on doing easy questions.

3)   They don't believe in their ability to learn. Once they got their answer wrong, it is hard to persuade them to do correction because they think they would not be able to do such questions by themselves in exams. The whole point of learning is to learn something that we were not able to do. We must reflect to see how far we have come in learning things that we did not know before. If we are not confident, we must practice until we are confident and not run away from such problems.

4)   They do not want to write down what they have understood and are too eager to move on. They put their priorities in learning wrong. Doing correction is more important than doing new questions. Penning down what you have understood is as important as understanding, if not more important. They think they will remember but do not make sure they will have

reference. They do not try to guarantee their learning by putting in place penned down corrections that they can use for reference later.

5) When I give tuition to students, some will immediately make corrections to every step in the workings that are affected by the mistake the moment I point out their mistake. Some will immediately redo the whole question. Both are fine. But there are those who just want to move on to the next question or need me to repeatedly prompt them to do correction; they don't have the habit and initiative to do correction; this last type of student will usually perform poorly.

MEMORIZE FORMULAS

In some system of learning, students are given formula sheets during the exam so they can focus more on the explorative or investigative type of learning. In these types of systems, the exam questions are usually easier. I can tell you this because I tutor a lot of students from the IB SL and HL syllabus as well as Cambridge Singapore A level H1 and H2 math.

My personal view is that memory and analysis aid each other. If I put a question requiring knowledge of a few formulas to students of both systems, the one that trains the student to remember formula will be able to do it readily, but the one that does not would have no clue of how to go ahead with the question. Even if the formula is in the formula sheet ready for reference, but the student who did not memorize the formula does not have anything to associate or process. How do you think about something if in your mind it does not exist any related facts and knowledge? Arguably, a lot of knowledge is out there in Google for the real-world purpose; but then students should be able to bring a laptop into the exam hall, but that is not the case yet.

Do we need to have a strong memory to perform an intellectual task in real life? Will we be faster and more effective if we have a strong memory? I would think so base on my life experience even beyond academia. Therefore, I see it is worth it to train up my memory by memorizing formulas.

If you are studying in a system that provides you with a long list of formulas for you to bring into the exam hall, I suggest you still memorize all the formulas for the purpose of training up your memory and you will find that you will be able to do more difficult questions than your peers who do not try to memorize the formulas.

To be fair, time is an opportunity cost. You can spend it on memorizing formulas or doing something else. But let me promise you, if you memorize the formula, you will be able to do better in exams and have stronger memory and will be able to memorize things faster as you go.

As for other considerations, there will be valid arguments from both sides, it is what you want in the end, a stronger memory or other things. But to do well in math exams, I highly recommend you memorize formulas, concepts, and methods.

A few ways to commit concepts, formulas to memory:

1)   Practicing the same type of questions again and again until you can never forget the formula or method.

2)   Cover the written formula, picture it in your mind, open up the written formula to check. After some time, picture it in your mind again and check with the source to ensure you remember correctly. When you can recall the formula effortlessly and accurately without fail even after a time gap of a few weeks or month, you have successfully committed the formula into long term memory.

3)   Observe any pattern you can in several formulas and try your best to understand how they are derived. Some formulas look alike, create stories to help you memorize them together yet distinguish clearly between them.

In the subsequent chapters, I will highlight the top 10 Pitfalls that I observed of students studying mathematics and the methods to overcome those Pitfalls.

PITFALL 1 – MEMORIZING FORMULAS JUST BEFORE THE EXAM.

Problems with that:

1)   You don't have enough time, brain energy or peace of mind to do that effectively right before an exam. Don't forget formulas are just the very basic, the student still need to spend time learning concepts and how to apply them. So if you spend your time before the exam memorizing formula, how do you have time to revise the concepts and try the applications.

2)   Most likely you will remember some formulas wrongly and because they are not committed to long term memory, it will take a longer time and more brain energy to recall during the exam, making it more likely for you to blank out or panic during the exam itself.

3)   You are likely to forget the formulas after the exam, which means that you will snowball the number

of formulas you need to remember for the next exam, which will involve more topics.

How to overcome this pitfall?

Memorize the formulas on the day you are first shown those formulas. Challenge yourself to memorize everything shown to you. Challenge yourself to recall them.

PITFALL 2 – NOT PLANNING OUT YOUR REVISION TIMELINE, NOT BEING CONSISTENT IN YOUR STUDIES.

Consistency in studies means making a habit to complete all assignments given by your teacher. If your teacher does not give an assignment or is ineffective in teacher, you should buy a recommended assessment book or textbook that provides examples, solutions, and questions and do them on your own following a regular pace.

You also need to do regular revision on topics that have not been touched for a month or more because the brain has the habit of pruning away memories that are no longer useful, but if you revise regularly, the formulas and concepts will form part of your long term memory.

The dates for exams are usually announced more than a month before. You should plan out your revision based on the number of days you have.

For example, If there are 21 days to your exams and 8 topics you need to revise;
then you should plan out your revision timeline to cover most topics within 3 days and 1 easy topic using just 2 days.

Every day you let yourself procrastinate means you have more to cover in the later days, which means you are going to be less effective.

If you have been consistent in your studies, you should use the strategy of doing 1-2 questions from each of the eight topics every day to keep every topic fresh in your mind.

If you have not, then you may want to conquer topic by topic. Though less effective, but you have no choice because you lack the fundamentals to the first strategy.

PITFALL 3- GIVING UP TOO QUICKLY

When you see a math question that is challenging, if you give up within the first 4 hours of working on it, that is giving up too quickly.

As I mentioned, I can spend days or even months working on a single question. But it is not staring at the question for 4 hours. This is what you need to do within the first four hours.

1)   Try all the methods and concepts you think possible.

2)   Flip textbooks for similar questions and study the solutions to them.

3)   Google online using the key concept words of the question. Especially the words related to a mathematical concept.

4)   Work out your own solution, check for careless mistakes, check with the answer provided.

5)   After you have worked out the solution if you are not sure that you can do it without reference to a textbook or other source, redo the question again without referring to anything or find a similar question to work on.

PITFALL 4- DECIDING WHETHER IT IS A WASTE OF TIME WHILE DOING IT.

If you have any doubt about doing what you are doing, then you should leave it and do something else that you have no doubt about doing. The worst thing to do is to keep deciding whether something is worthwhile doing it. It slows down your progress a lot and wastes your time. If you have been wishy-washy about something for more than 3 times, decide once and for all that you are going to quit it or if you know you cannot quit it, just give all your best to it and do not entertain the wishy-washy thoughts again.

PITFALL 5 – AMBIGUOUS NOTATIONS, UGLY HANDWRITINGS

If you have practiced enough questions, you will naturally know how some ambiguous notations and ugly handwritings can lead you to make mistakes.

There is no ambiguity in mathematics when you use the correct notations, the problem is when individual students do not follow the correct way of denoting their workings.

Some students do not have the habit of labeling clearly what they are finding. It makes it hard for them to search for information that they need or for the teacher to follow their thought process.

For some students, the 5 is written like 3. Symbol of multiplication is confused with x. If x is used in an equation, it is better not to use the multiplication symbol because they look alike. The student should use 3(4x) instead of 3x4x. Subscripts and superscript should be written clearly. If there is not enough space between the lines given, then use two lines.

The principle is: if it is possible for two interpretation of the notation then it is an ambiguous notation, the student should make effort to find out what is the correct way of denoting and not just invent their own way. He needs to be careful and check against a proper reference. A peer is not a good reference and he is also learning unless he is a top scorer who gets 90 marks or above for his exams, one should learn from textbooks or expert teachers. Which brings me to my next point.

PITFALL 6-LEARNING FROM CONFUSED PEERS

When I'm teaching classes, I realize students often learn from their peers who are clueless or teach the wrong things. As a result, the wrong concepts are spread in the class like wildfire in a forest. If you learn the wrong things, it forms an impression which you may erroneously recall, especially if you are not the hardworking or careful type. Therefore, it is important to always refer to good textbooks and teachers who can explain the concepts clearly. It is better to think by yourself carefully on a problem and referring to textbooks than to ask a peer and accept his/her method without proper thinking and reasoning.

However, there is an exception to this if a peer has demonstrated strong ability in mathematics.

PITFALL 7 – SATISFACTION WITH GETTING AN ANSWER

Students should not be satisfied with getting an answer to a question, especially if they did not get the answer by themselves.

If someone helps you get the answer, you should try to work it out by yourself again to see if you can get it without help.

Secondly, even if you got an answer to a question by yourself, you should still try to address doubts that arise in the middle. For example,

"I remember seeing a similar question, could I do that question too?"

"I don't remember having taught this method, is it not in the lecture notes?"

"I should write this concept down somewhere for easy reference when I need to do revision."

"How do I remember this so that I won't forget."

"Why is the solution showing a different method, I should try to understand the method too and compare their effectiveness."

"What are the key concepts and steps to this question, which ones are harder to grasp that I should pay more attention to?"

PITFALL 8-NOT SETTLING WITH A SET OF STUDY HABITS THAT WORK FOR YOU.

When I was a student, I would lie on the bed to think about a question for hours. I would switch on the TV and watch while I do revision. These habits help me to persist in my studies for hours because firstly there is no strain on my back from sitting for too long and I can concentrate better on the bed. I would put all the reference materials and stationaries I need right beside me, so I don't need to be interrupted.

However, don't presume that the habits that work for me will work for you.

It is important to settle on a habit that will allow you to focus for hours. Some students go to the library and put on their earphone playing the music that doesn't distract them so as to block out the noise from the environment.

Do the things that work for you and don't keep changing your study habits if you already found one that works.

PITFALL 9 –NOT LOOKING AT THE QUESTION WHEN THINKING.

You should be looking at the question or textbooks when you think. Maybe you have not gotten some critical information from the question or the textbook. It does not help to look away and just think when you have not understood the question fully.

When teaching students, often I find students looking away from the question too quickly to think when it is obvious to me that they have not registered a piece of critical information from the question. If so, it is pointless to start thinking.

PITFALL 10 -NOT REWRITING THE QUESTION IN A FORM THAT IS EASILY UNDERSTOOD BY YOU.

Some questions are wordy, after reading you may not even understand them or what they are asking. In such situation draw some diagrams and input all the information into the diagram for the mind to process them visually. Rewrite sentences into mathematical equations. Use algebra to represent the unknowns that you are trying to find.

NOT FILING YOUR NOTES, TUTORIALS PROPERLY
Homework can be given to students in a loose sheet of paper. Students need to be consistent in their effort to file them neatly.

It is best to file lecture notes, tutorials, exams and test papers separately. They should be ordered in terms of topics. A ring file where you can open to insert or extract sheets in between would be convenient.

The benefits of filing your notes properly are as follow:

1)   Easy retrieval for revision.

2)   A systematic filing helps in systematic revision.

3)   Saves time and energy from finding your notes among a pile, which can be demotivating. Imagine if you are doing a question halfway and need a piece of information or reference but you know it is somewhere in a big messy heap, you will likely give up on that idea of finding that reference. Worse still if you lose them.

PITFALL 11 – SLEEPING LATE BEFORE THE EXAM DAY & STUDYING INTENSIVELY BEFORE THE EXAM

Panic attack or blank out during the exam are mostly caused by these 2 bad habits.

Doing math requires a lot of brain power. You need a minimum of 7.5 hours of sleep the night before; it matters that you sleep before 11pm on the day before the exam, get yourself as relaxed and comfortable before the exam. If you have not studied consistently, studying in the hours right before the exam does not help but is more likely to worsen your performance during the exam by draining your brain power.

PITFALL 12 – BAD EXAM TACTICS

1. Overdoing on presentation leaving no time to work on the other questions.

   How to overcome: You should complete all the questions first, before going back to polish up on the presentation.

2. Spending too much time on difficult questions.

   How to overcome: You should separate exam questions into 3 rounds. In the first round you should do only the easy questions, those that you can do with relatively little thinking. Mark a tick on the questions that you have completed.

   You can still work in order from the 1st question to the last one for the first round but skip whenever you find a question hard. Make a mark on the questions that you skip. Mark 2 if it is a question that you think is doable with a bit more thinking. Mark 3 if it is a question that you think you will need to think for a very long time.

3. The pitfall of poor handwriting and ambiguous notation has already been mentioned earlier.

But I will like to give some additional examples and tips here:

Tip 1: It is more helpful to use B to represent the cost of a banana than to use Y even though both representations works. However, it lessens the load of the brain in a complex question when you use B since B can associate with banana more easily than Y by the first alphabet of the word.

Tip 2: Label all that you found. Some students don't have the habit of labelling.
They would write
$3(20+4)/10 = 7.2$
when they it is clearer to label:

cost of banana = 3(20+4)/10=7.2

Labelling is especially important in a question with many parts. Not only will it help the marker better understand what you are doing, it also helps you check your own work or find information that you worked out if you need them to work out other things in the question.

Not labelling what you are finding adds the complexity of reading your own workings to the complexity of the question you are working on, making it harder to think for a solution.

Tip 3: Use tables and diagrams to simplify your presentation of information so that you can think   more   clearly.

First example: A wordy question

A solid with height 7cm has a cross section as a triangle with sides 4cm and 0.02m. The angle between the side of 0.02m and the other side with unknown length is 30 degrees. Find the volume   of   the   solid.

Try representing all the above information in a diagram and have all the information written down in the diagram itself. It will make it easier to work on the question.

2nd example:

Which takes less time to write but is easier to read? I would think the table is usually a better way to represent, especially when the number of entries get larger.

$P(x = 3) = 0.2$
$P(x = 2) = 0.4$
$P(x = 1) = 0.4$

a	P(X=a)
1	0.2
2	0.4
3	0.4

CONCLUSION

How to use this book?

You can read this book when you are not sure why you are not doing well in mathematics. You can use this book to identify the possible problems with your study approach and correct them.

If you have a question, feedback or comment, please email to genie10001@gmail.com or WhatsApp 81808611, I'm happy to guide you or learn from you.

You can find out more about my home tuition services at www.h2math.com